Rocket Rangers

Man's Quest to Fly Like Buck Rogers

Aerospace Heroes Commemorative

Vol. 2

With vivid photo histories of the
Jetvest, Rocketbelt, Jetbelt, MMU, SPK and S.A.F.E.R

Edited by:

Nelson Louis Olivo

Aerospace Historian

This book was printed in the United States of America.

To Order More Books
The Manhattan Literary Agency
1-718-403-0256
the Xlibris Bookstore at www.Xlibris.com
Amazon.com
Barnesandnoble.com

Summary: A vivid photo history of man in free flight from the days
of Buck Rogers and the early rocketbelts to the era of the MMU.

For

Margarita Pagan – my Mom

And in memory of our loved ones and friends

Judy Wells, Tom Lennon, Thomas Moore, Gordon Yaeger,
Wendell Moore, Robert Courter, Harold Graham

Dedicated to

the Young Rocket Rangers

the next generation of dreamers and doers

New Jersey's Young Rocket Rangers

Andre Castro, Savien Castro, Lesly Mia Sapón Gonzalez,
Omar Castro, Deidy Marisol Sapón Gonzalez,
Miriam Guadalupe Lopez Gonzalez.

The dream of every space-bound Young Rocket Ranger

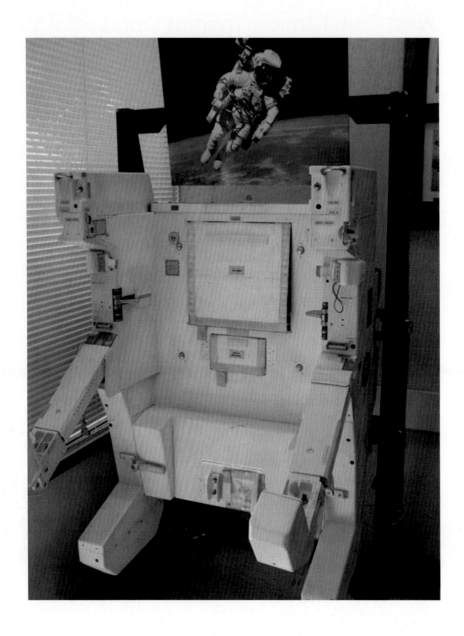

to fly The MMU

Contents Volume 2

The First Rocket Ranger Astronauts   73
Yuri Gagarin, Alan Shepard

Astronaut Maneuvering Units   75
Wendell F. Moore's Zero Gravity Belt,
The Lunar Rocketbelt & Dual Maneuvering Unit.
The Air Force Astronaut Maneuvering Unit
Ed White & The Hand-Held Maneuvering Unit.

NASA's Mission Control Center   81
Alan Glines: NASA's Mission Control Center &
      Apollo 13: From Buck Roger's Science Fiction
      to Apollo Science Fact.

Skylab and Astronaut Maneuvering   92
Alan Glines: The Astronaut Stabilized Maneuvering Unit,
      The Modified Hand Held Maneuvering Unit,
      The Jet Shoes.

Flying Like Buck Rogers   95
The American MMU
Astronauts Bruce "Buck" McCandless II &
Robert "Flash" Stewart go Flying like Buck Rogers.
George "Pinky" Nelson & The Solar Max rescue.
James "Ox" Van Hoften & his Cargo Bay Rocket Ride.
Joseph Allen & the Palapa rescue.
Dale Gardner's "Stepping Out for a Walk."

The Soviet MMU-SPK   135
Cosmonauts Alexander Victorenko and Alexander Serebrov
go flying like Buck Rogers.

The S.A.F.E.R.   142
Mark Lee and the new S.A.F.E.R – the Jetvest styled mini MMU test.

End Page   148
Harold "Pete" Graham

Putting Your Dreams to Flight:
Training   149
Aviation Careers   152
Study Resources   154

Nelson's Inspiration   161

By 1930, Buck's sci-fi comic book world of the future had captivated millions and while his devoted young fans dreamed of one day flying like Buck Rogers, adults would never believe that rocketships much less rocketmen, could or would ever exist outside of the monthly comics.

Were they right?

No!

Within a short span of 25 years aeronautical engineers would begin to seriously look at the impossible and dream those very same dreams!

While many thought this 1929 comic's illustration was just a daydream, it accurately predicted how future astronauts tethered to their capsule and using HHMU's would float in space.

On April 12th, 1961 those Buck Rogers dreams took on real life as Yuri Gagarin became the first human to rocket-fly above the earth's atmosphere.

Yuri Gagarin the 1st Cosmonaut
Alan B. Shepard the 1st Astronaut

On May 5th, 1961 Alan B. Shepard's flight made him the first Astronaut.

Following closely in their footsteps were Astronauts Gus Grissom (July 21, 1961), Cosmonaut Gherman Titov (August 6th, 1961) and Astronaut John Glenn Jr. (February 20th, 1962).

Cosmonaut Alexi Leonov

Then when Cosmonaut Alexi Leonov boldly stepped outside of his capsule on March 16th, 1963 with just a thin tether between him and his ship, an astronaut-maneuvering unit like the one worn by Buck Rogers was the next logical step.

Anticipating the eventual need for an orbital flight repair, in December 1960 the aerospace medical division of the U.S. Air Force issued a report entitled "Self Maneuvering for the Orbital Worker." This report outlined the needs an astronaut would have when he stepped outside of his craft to execute an EVA (extra-vehicular activity).

One of the first to design an astronaut-maneuvering unit during this early space period was rocketbelt creator Wendell F. Moore. His first operational prototype was called the Zero Gravity Maneuvering Belt (ZGMB). Worn around the waist this silver-metallic snap-on belt was designed to provide thrust assistance to the outer space worker.

Zero Gravity Maneuvering Belt Tested

Air Force test officer rises to the cabin ceiling in an Air Force C-131 during zero gravity test of Bell Aerosystems Co. Zero Gravity Belt. Belt is designed to allow occupants of manned space vehicles to have a means of maneuvering when they leave the spacecraft for inspection, repair or assembly tasks. Nitrogen gas under pressure has 20 lb. thrust.

For flying across lunar surfaces, Wendell designed a rocket-powered astro-chair he called a Lunar Rocketbelt (LR). Wendell also created the astro-pogo, a stand-up Single or Dual Maneuvering Unit (SMU, DMU), for air, space, and lunar transportation.

Astro Chair	Astro Pogo

Aerotest pilot Robert Courter tests the chair and pogo

In 1961, the U.S. Air Force requested bids for a study on EVA maneuvering units; Ling-Temco-Vought (LTV), an aerospace systems company was selected. The study led to a design of a buck roger's propulsion backpack called the Self Maneuvering Unit (SMU). LTV then received the commission for an operational model of the SMU which LTV renamed the Astronaut Maneuvering Unit (AMU). Years of ground tests followed and finally in 1966, the AMU was scheduled aboard flights 9 and 12 of America's second manned space program called *Gemini*.

The world would not have to wait that long however, to watch the first American Astronaut EVA or the first space test of an astronaut maneuvering unit. In June of 1965 Astronaut Edward White took a 22-minute rocket-assisted float outside his Gemini 4 spacecraft. To maneuver about in space, Edward used a small Hand-held Maneuvering Unit (HHMU), also known as the Zip Gun, which was made from Gemini ejection seat oxygen bottles, a pressure regulator from a Mercury environmental control unit, and a control handle with three thrusters attached.

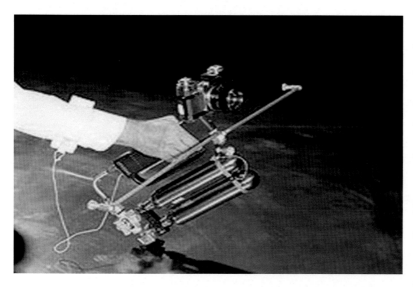

By squeezing the forward end of the handle, he operated the tractor thrusters and moved backwards. By squeezing the aft end of the handle, he operated the pusher thrusters and moved forward. A life support umbilical provided breathing air.

With the aid of his HHMU, Edward exited his capsule, traveled the length of the craft and performed a few pitch and yaw maneuvers, leaving behind unforgettable pictures of his venture into the void. Another first was that Edward's EVA was broadcast live. Every thrilling moment of his flight took each and every viewer one step closer to the future, a future first forecast in the comic strips of Buck Rogers.

Astronaut Ed White floats in space

Astronaut Ed White

The HHMU was scheduled on board the flight of Gemini 8, but the mission was aborted before an EVA could be achieved. On the Gemini 10 flight in July 1966 Astronaut Michael Collins used the HHMU. The unit was identical to Edward White's except that it was fueled by nitrogen contained in a storage bottle in the spacecraft. A hose attached to the life support umbilical carried pressurized gas to the unit. The task was to retrieve a micrometeoroid package mounted near the capsule's nose.

Using his HHMU Michael tried to fly over to the package. At first, he had trouble maneuvering and developed a pitch-down rotation. Attempting to correct his balance, he found himself translating upward. On a second attempt, he retrieved the package and completed the task. It was now the AMU's turn to be tested.

Astronaut Eugene Cernan planned to fly around his spacecraft for an hour connected to the Gemini 9 only by a 125 foot safety tether.

Drawing of extravehicular activity showing the Astronaut Maneuvering Unit (AMU) which was to be worn by astronaut Eugene Cernan during the Gemini 9 mission in 1966. Use of the device was aborted when Cernan's space suit cooling system was overloaded. (NASA)

The AMU had a life support system built in so a life support umbilical as used with the HHMU wasn't needed. The AMU's twelve thrusters were powered by hydrogen peroxide using the same silver catalyst principle as the rocketbelt. A cover layer made of Chromel 9-R stainless steel fabric protected Eugene's legs from the exhaust plume. In struggling to don his AMU, Eugene overloaded the spacesuit's cooling system. As had happened many times before, the EVA had to be postponed.

Astronaut Dick Gordon was scheduled to use an HHMU similar to Collin's during the flight of Gemini 11. Like Eugene, Dick also had to curtail his EVA because of an overheated spacesuit. The numerous AMU/HHMU complications caused the testing of maneuvering units to be tabled. Not until the arrival of *SKYLAB* did maneuvering unit tests restart.

Astronauts Michael Collins, Eugene Cernan and Dick Gordon

NASA's Mission Control Center
(as recalled by former Apollo Flight Controller Alan Glines)

Readers note:
Behind these doors are the men of Mission Control;
men who fly like Buck Rogers without leaving the Earth.

At the dawn of the space age, specialists from a variety of fields were called upon to apply their knowledge to the new and emerging field of aerospace. To house these specialists, in 1965 the creation of a spaceflight control center began at Cape Canaveral, Florida.

Flight specialists dutifully monitored every aspect of each mission from the start of the countdown to the moment the astronaut was safely recovered from his capsule's splashdown, or as with today's Shuttle, the landing gear wheels rolled to a stop.

The flight specialists soon came to be known as flight controllers and the spaceflight control center soon came to be known as the "Mission Control."

Today's Mission Control Center (MCC) is as modern and scientific as any used by science fiction's Buck Rogers. Based at the Johnson Space Center (JSC) in Houston Texas, the Mission Control Center supports every facet of the manned space programs and is operated by NASA (the National Aeronautics and Space Administration).

The new Mission Operations Control Room (MOCR) within Mission Control is the operational nerve center. It is from this room that the needs of all the present day Rocket Rangers are met.

The MOCR illustration that follows identifies the Flight Controllers and their call signs.

FC # 1. Public Affairs Officer (PAO) provides mission commentary to supplement and explain air-to-ground voice transmissions and flight control operations to the news media and the public.

FC # 2. Mission Operations Directorate Manager (MOD) provides a link from the MOCR team to top NASA and JSC Missions Operations Directorate management personnel.

FC # 3. Russian Interface Operator (RIO). The Russian Interface Officer serves as the primary interface between the U.S. and Russian control teams.

FC # 4. Surgeon (Surgeon), monitors crew activities and health status.

FC # 5. Integrated Communications Officer (INCO), plans and monitors in-flight communications and instrumentation systems configuration including onboard television.

FC # 6. Flight Director (FD), call sign "Flight," serves as leader of the flight control team, and is responsible for overall Shuttle mission and payload operations, and all decisions regarding safe, successful flight conduct.

FC # 7. Spacecraft Communicator (CAPCOM) call sign "Capcom," serves as primary communicator between flight control and astronauts. The initials are a holdover from earlier manned flight, when Mercury was called a capsule rather than a spacecraft.

FC # 8. Payload Deploy Retrieval (PDR), monitors operation of the remote manipulator system.

FC # 9. Data Processing System Engineer (DPSE) is responsible for data processing system including the five onboard general-purpose computers.

FC # 10. Payloads Officer (Payload) coordinates onboard and ground system interfaces between the flight control team and payload user.

FC # 11. Flight Activities Officer (FAO), plans and supports crew activities, checklists, procedures, and schedules.

FC # 12. Electrical, Environmental, Consumables Manager (EECOM), responsible for environmental, air, and water resources.

FC # 13. Propulsion Engineer (PROP) monitors and evaluates reaction control and orbital maneuvering propellants and other consumables available for maneuvers.

FC # 14. Guidance, Navigation, and Controls Systems Engineer (GNC), monitors all vehicle guidance, navigation, and control systems.

FC # 15. Maintenance, Mechanical, Arm, and Crew Systems (MMACS), call sign "Max", monitors operation of the orbiter's structural and mechanical system.

FC # 16. Electrical Generation and Illumination Engineer (EGIL), monitors electrical systems, fuel cells, and associated cryogenics.

FC # 17. Flight Dynamics Officer (FDO), call sign "Fido," plans all maneuvers and is responsible for the overall trajectory from launch, on-orbit operations; de-orbit, entry, and landing.

FC # 18. Rendezvous (RNDZ), flight dynamics controllers responsible for rendezvous trajectories.

FC # 19. Ground Controller (GC), ensures the MCC is functioning properly and coordinates outside data and communications traffic.

20. World map Screen.
21. TV Screen.
22. Mission Clocks/Telemetry Data

The Houston Mission Control Center as it appears today

One of the most valued visual support systems are the displays, a series of projection screens on the front wall of the Mission Operations Control Room. The displays range from plotting charts that show the spacecraft's location, to actual television pictures of activities inside the Shuttle, as well as views of Earth, payload deployment/ retrieval, and EVA's.

Today, 19 flight controllers staff the Mission Operations Control Room but it takes on the average 50 MCC members working nine-hour shifts to fully monitor a spaceflight. Leading each 50-member team is a flight director who is supported by a CAPCOM.

Note: The new MOCR for current Space Shuttle and Space Station missions is located on the 2nd floor of JSC's Building 30.

The original Apollo Flight Control Room, the one that I was assigned to is on the 3rd floor and is a National Monument preserved for history.

The Historic Apollo Flight Control Center

My interest in outer space began when, as a young teen I started reading the comic strip Buck Rogers. A few years later I can remember being glued to the radio in my high school classroom while my classmates whispered the countdown (5-4-3-2-1- IGNITION) to Astronaut Alan Shepard's liftoff. I also remember how proud we students were of America's first man in space.

I went on to college; graduated with a Bachelor in Electrical Engineering and with my good grades in hand I applied to NASA for a job. The value of good grades revealed itself when not 1, but 16 job offers arrived from the many divisions at NASA. In 1966 with an eye on the final frontier, I chose the offer from the new Mission Control Center in Houston.

It took years of training but I eventually became one of the five senior controller's responsible for the operation and control of all onboard Apollo and Skylab spacecraft communication systems. "INCO" was my call sign in the main control room, which, in our world of acronyms means Integrated Communications Officer (see FC#5).

The communication systems I was responsible for included all the voice, telemetry, tracking, television, and command systems on the spacecraft. Take TV control as an example. On Apollo, INCO was the operator of the small television camera on the front bumper of the Lunar Rover, controlling the pan, zoom, and tilt from the digital command push buttons on his console.

On Skylab, INCO recovered all the nine months of video that the three crews recorded through his control of the onboard video player. Each night, while the Skylab crew slept, INCO would play back the day's recorded video to tracking stations in the US, which relayed it back to Mission Control.

Apollo 13: From Buck Rogers Science Fiction to Apollo Science Fact
(as recalled by former Apollo Mission Controller Alan Glines)

Flight controllers at NASA learn to anticipate logically almost all possible occurrences. However, there is always that rare element of chance in every mission. A case in point is Apollo 13.

On April 11th, 1970, Apollo 13 lifted off for the Moon with Commander Jim Lovell, Command Module Pilot Jack Swigert and Lunar Module Pilot Fred Haise aboard. Two days later, with the spacecraft nearly at the Moon and 200,000 miles from Earth, an oxygen tank EXPLODED, scrubbing the lunar landing and putting the crew in serious jeopardy. "Houston, we have a problem," were the now famous words from Apollo 13 Commander Jim Lovell that sent our operations team into hyperdrive and stretched our expertise to the limits.

Those first nerve-wracking hours are imbedded in my soul. Working hand-in-hand with us the crew successfully turned their 2-man lunar module into a 3-man "lifeboat," and even rigged an adapter so that a command module "air scrubber" would work in the lunar module, preventing a lethal buildup of carbon dioxide. Then 2 critical days later came the harrowing second-by-second re-entry which had us all in silent prayer. When on April 17, 1970 Apollo 13's parachutes opened and it safely splashed down, one could hear our CHEERS across the street, across the city, across the state!

INCO

Apollo Mission Controller Alan Glines at his "INCO" station during the Apollo missions.

Reader's Note:

In recognition for their miraculous Apollo life-saving efforts, Alan Glines and each member of the MCC operations team was presented with the coveted Presidential Medal of Freedom, our nation's highest civilian honor.

PRESENTED TO:
ALAN C. GLINES
OF THE
MISSION OPERATIONS TEAM

The President of the United States of America

Awards this

Presidential Medal of Freedom

To

The Apollo XIII Mission Operations Team

We often speak of scientific 'miracles'—forgetting that these are not miraculous happenings at all, but rather the product of hard work, long hours and disciplined intelligence.

The men and women of the Apollo XIII mission operations team performed such a miracle, transforming potential tragedy into one of the most dramatic rescues of all time. Years of intense preparation made this rescue possible. The skill, coordination and performance under pressure of the mission operations team made it happen. Three brave astronauts are alive and on Earth because of their dedication, and because at the critical moments the people of that team were wise enough and self-possessed enough to make the right decisions. Their extraordinary feat is a tribute to man's ingenuity, to his resourcefulness and to his courage.

The White House
Washington D.C.
April 18, 1970

SkyLab and Astronaut Maneuvering
(as recalled by former Apollo Flight Controller Alan Glines)

SkyLab, America's first space station was launched on May 14, 1973. It was four stories high, constructed from the Apollo Saturn 5 moon rocket stage, named the S-IV-B. It flew at an altitude of 270 miles above the earth and orbited the Earth for 9 months. Skylab's primary mission was to determine if people could survive in outer space for long periods.

Skylab

Skylab also allowed for many orbital experiments, which included the testing of maneuvering units. Scheduled for testing aboard Skylab were: (1) the Astronaut Stabilized Maneuvering Unit (ASMU), a backpack predecessor to the MMU which tested successfully, (2) NASA's own HHMU with modifications and my all-time favorite, (3) the Jet Shoes, whose cartoon replicas were flown by the Jetsons on television

New Hand-Held Maneuvering Unit   Astronaut Gerald P. Carr flies the ASMU

Jet Shoes

Unlike the hand-controlled AMU or HHMU the Jet Shoes were foot control maneuvering units with eight thrusters on each foot. The Astronaut could rotate about on all three axes but could only translate along the up and down axis and not side-to-side.

Thrusters were mounted on stirrup-like foot holders attached to the mainframe or saddle. The nitrogen gas propellant tank was mounted to a back frame and worn like a backpack.

To spacewalk with the Jet Shoes, the Astronaut moved his feet and toes. Though never used in any space program, the idea behind the Jet Shoes was to free the Astronauts hands for other extravehicular activities.

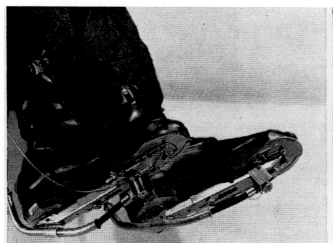 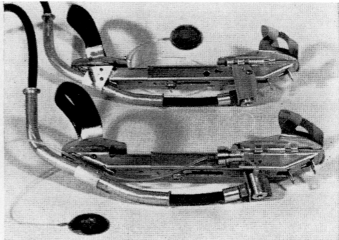

Flying like Buck Rogers

The American MMU

Major C.E. Witset, Cliff Hess, and Bruce McCandless II designed and developed the Manned Maneuvering Unit (MMU) or as it's more popularly known with the fans of Buck Rogers, the Jetpack. The MMU was the first rocket-powered backpack designed to provide a full array of flight support systems as well as controlled maneuvering capabilities outside a spacecraft in zero gravity.

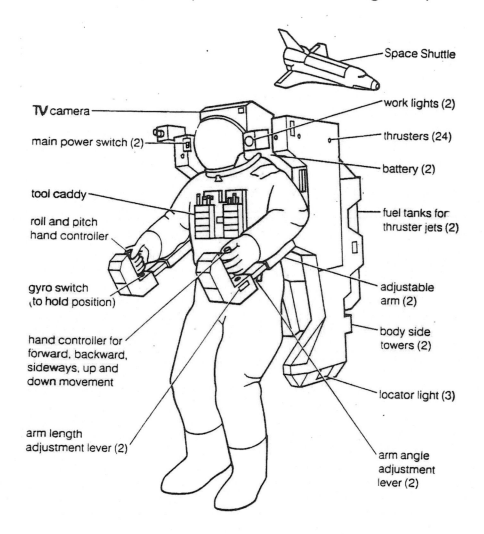

American Manned Maneuvering Unit

The first MMU to be flown in space was the result of combining imaginative daring, creative design, and sophisticated technology.

The tank valves and switches were off-the-shelf components, all of which were tested on unmanned spacecraft. Parts were adapted from the Viking Mars Lander and other free-flying craft. The propellant tanks came from a helicopter; other propulsion parts came from the Space Shuttle. The parts then had to be packaged to fit through the Shuttle hatches that connected the payload bay with the crew cabin through the airlock; this allowed the MMU to pass through into the cabin for repairs, if needed.

Astronaut Bruce McCandless II models the new MMU

Building the MMU also required maintaining a center of gravity (a balance of both weight and controls) for the man-machine combination; all the operational devices (pressure gages, toggle switches, power switches, locking latches) had to be conveniently placed. NASA began the development of an easily controllable maneuvering unit in 1974 and conducted experiments until October 1983.

After nearly a decade of development, the MMU had become what Astronaut Joseph Allen called "a three-dimensional flying carpet," allowing the pilot to move about effortlessly in space.

To fly the MMU an astronaut uses hand controls mounted on the armrests. The left hand determines speed and direction while the right hand determines attitude or in pilots terms, pitch, yaw and roll. Equipped with an MMU an astronaut literally becomes a human spaceship with his eyes as his optics, his hands as his manipulators, and his brain as his computer.

The MMU's operational speed is 1/3 to 1 mile per hour, with a top speed of approximately 40 mph. And it can function for six hours untethered between charging, an advantage over the hand-held propulsion units, or Zip Guns as they were known, which emptied quickly and were umbilically tied to the space capsule.

For extra safety, the MMU was given a dual parallel operating system. If one system failed, the astronaut would use the backup operating system to return safely to the Shuttle cargo bay.

Bruce McCandless II tests an experimental MMU

Designed to work in balance with the human physique, the MMU's dimensions are 127 centimeters high, (50 inches) by 84.6 centimeters (33.3 inches) wide, by 68.6 centimeters (27 inches) deep with the arms folded, 122 centimeters (14 inches) deep with the hand controller arms fully extended for flight; a little over four feet tall.

On Earth, the MMU weighs 338 pounds loaded, including 26 pounds of nitrogen propellant in two canisters each pressurized to 300 psi; in Space it weighs zero lbs. Maneuverability is maintained by its 24 fixed-position gaseous nitrogen thrusters. Depending on the crew member, the operational weight of the MMU is between 640 and 765 pounds.

Scheduled aboard the first Shuttle mission in 1980, the MMU's flight plans allowed for an astronaut EVA (Extravehicular Activity) to inspect and repair any damaged Shuttle thermal protection tiles.

But fearing the MMU might fail, NASA cancelled the EVA. It was the backup safety factors of the unit that eventually convinced NASA to let astronaut/co-designer Bruce McCandless II perform the first history-making, solo-flying, rocket-powered spacewalk.

Astronaut Bruce McCandless II
goes Flying Like Buck Rogers
(with historical in-flight commentary)

Finally, in the spring of 1984, Challenger orbiting over the center of the earth slowly opened its payload bay doors. Astronaut Bruce McCandless II was waiting inside with his MMU. On board the shuttle and prior to his walk, Bruce had become affectionately known as "Buck," (Rogers) and now he was going out to live the part.

Once flight permission was given Bruce pulled a small handle on the armrest of his MMU and fired the thrusters. Gently he rose head-first out of the Challenger into the vast expanse of space.

Bruce McCandless II lifts up and out of the shuttle's cargo bay

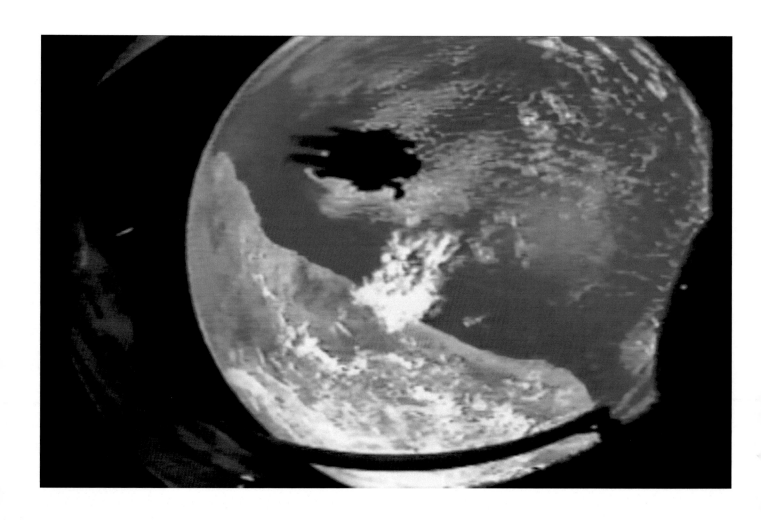

Bruce McCandless II glides away from the shuttle

Twisting another knob, Bruce flipped towards the Earth and recognized the terrain below: "Looks like Florida—it is Florida! It's the Cape," Bruce said.

What had only been a comic book dream was now a living reality. Bruce McCandless II was circumnavigating the globe, 17,000 miles per hour and at 150 miles above the Earth! Looking down on the big blue marble we call home, Bruce exclaimed his now famous words: "It may have been one small step for Neil (Armstrong), but it's a heck of a leap for me!"

With a perfect first spacewalk achieved, Bruce maneuvered back to the Shuttle and hovered in the payload bay. Then a message from the White House arrived:

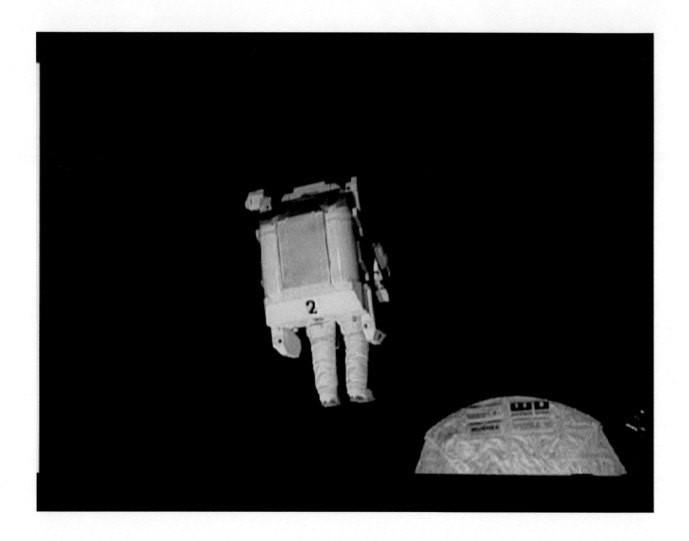

President Reagan making a telephone call to the space shuttle

"Let me ask you," said President Reagan, "what's it like to work out there, unattached to the shuttle and maneuvering freely in space?" Bruce replied: "Well, we've had a great deal of training, sir, so it feels quite comfortable. The view is simply spectacular and panoramic. And we believe that maneuvering units are literally opening a new frontier in what man can do in space."

The maiden voyage of the MMU complete, Bruce removed his MMU and took his place inside the Challenger. To his crewmates, the controllers at Mission Control, the President of the United States, and to every young astronaut hopeful watching him circumnavigate the globe, Bruce McCandless II had become the living embodiment of our comic book hero.

Astronaut Bruce McCandless II

Alternating with Bruce "Buck" (Rogers) McCandless II in two spectacular spacewalks was Astronaut Robert L. Stewart affectionately known as "Flash" (Gordon).

Astronaut Robert L. Stewart flying like Buck Rogers

Astronaut Robert L. Stewart

Following the history-making spacewalk, the Space Shuttle Challenger was launched on yet another history-making mission.

George "Pinky" Nelson and the Solar Max Rescue
(as recalled by former Astronaut James "Ox" van Hoften)

Before flying the manned maneuvering unit (MMU) in space, astronauts train on a special flight simulator. Early in 1984, Astronaut George "Pinky" Nelson and I began MMU flight and satellite rescue simulation, combined with satellite repair training for mission STS 41-C: the in-orbit rescue and repair of the Solar Maximum Observatory Satellite (Solar Max). Solar Max was the first satellite specifically designed for in-flight service. In April of 1984, Pinky and I put our specific training to the test.

James "OX" van Hoften training for his EVA

George "Pinky" Nelson training for his satellite rescue

Astronaut George "Pinky" Nelson

On the third day of the mission April 8th, 1984, Pinky, propelled by his MMU and carrying a TPAD (Trunnion Pin Attachment Device), flawlessly maneuvered through space on a rendezvous with Solar Max. Closing in on the slowly revolving satellite, Pinky skillfully flew in between the solar panels maintaining an equal spin speed.

An undetected protruding insulation fastener near T-PAD's locking pin prevented a successful docking with the satellite and Pinky piloted his MMU back to the cargo bay. Though disappointed with the T-PAD's failure, Pinky was delighted with the MMU. "It flew as smooth as glass," Pinkly later recalled.

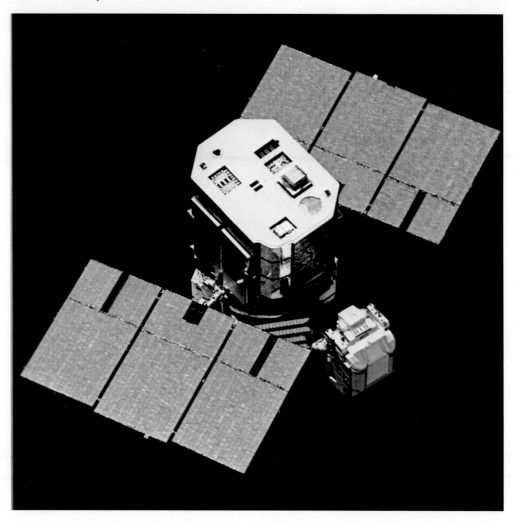

On April 11th , using the RMS (remote manipulator system) arm on the Space Shuttle, mission specialist Terry Hart grabbed Solar Max on the first try; seven hours later, Pinky and I completed repairs on Solar Max, and released it back into orbit. The first in-flight servicing of an orbiting satellite demonstrated the flexibility of the shuttle, the versatility of man, and the precision performance of the MMU. Mission 41-C was a history making first for the space program and an incredible adventure for the crew.

Then to cap it all off, I even got to don an MMU and do a solo test flight. Flying the MMU in the cargo bay I tested its ferrying ability within a confined area as well as its center-of-gravity offsets. Spacewalking in a MMU is the ultimate form of space mobility.

Astronaut James van Hoften goes for a rocket ride in
the Shuttle's cargo bay

Astronaut James "Ox" Van Hoften

The following November, the crew of Discovery set off on a second satellite rescue. Only this time there were two ailing orbiting satellites that needed retrieval. And unlike the stable Solar Max satellite that Astronaut Pinky Nelson set out to rescue, the Palapa and the Westar were both spinning!

Discovery soars upward towards the next mission

First to be rescued was the Palapa satellite. Astronaut Joseph P. Allen using his MMU flew over to it and using a "stinger," (a new version of Pinky's grappling/docking device) was able to safely attach himself to the satellite. But once attached to the spinning satellite he began spinning along with it.

Astronaut Joseph P. Allen rehearses capturing a spinning Satellite

Using the MMU's reverse thrusters, he was able to slowly stop his spinning and stabilize the satellite. Expanding the Shuttle's manipulator arm Astronaut Anna L. Fisher waited for Allen to fly alongside and dock so she could then pull the Palapa satellite into the payload bay. But complications arose and for nearly two hours Allen manually steadied the satellite as Astronaut Dale Gardner tried to berth it.

Through a remarkable combination of unrelenting persistence and precision flying in microgravity, Joseph and Dale finally positioned the satellite into the payload bay. Two days later, to capture the spinning ten ton Westar satellite, Dale donned his MMU, waited for the thumbs up and prepared for what seemed to be a very simple rescue.

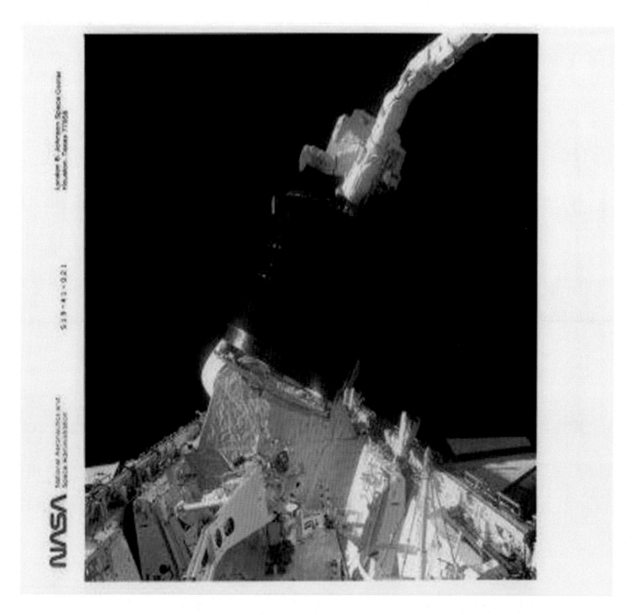

Astronaut Joseph P. Allen rescuing a satellite.

Astronaut Joseph P. Allen

Stepping Out for a Walk
(as recalled by former Astronaut Dale Gardner)

Despite the years that have passed since I flew the MMU from Discovery on Shuttle Mission STS-51A, the 14th Shuttle flight, my memories of that event are still vivid and lasting. I am often asked what it "felt" like to fly the MMU untethered from the Shuttle, becoming for a time a small human satellite in the vastness of space. It is memories of those feelings, as opposed to the technical aspects of the flight that are indeed the strongest to this day. A few of these I wish to relate because I believe they will be readily understood, even by someone who has not been in space, since they are based in part on the surprise and concern which each of us feels when facing the unexpected.

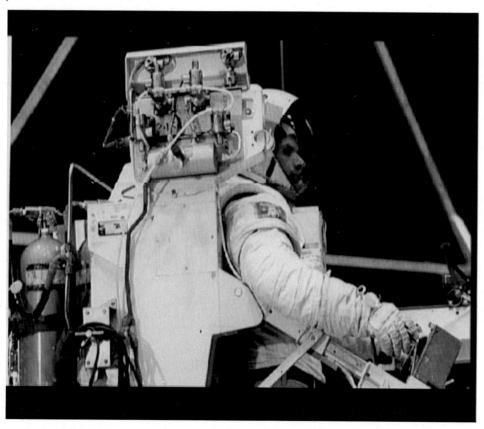

Astronaut Dale Gardner begins his satellite rescue testing

The first sensation that comes to mind from the flight is, not surprisingly, one from its very first moments. After donning the MMU in the payload bay of Discovery, I released the MMU from its support station; however, I was still attached to the shuttle via foot restraints that locked my spacesuit boots to a small pedestal. In that position I performed the final checks of the MMU's small nitrogen gas thrusters, then twisted my feet out of the foot restraints and gingerly "stepped" out into the openness of space.

I am sure the medical people in the Mission Control Center at Houston immediately saw a dramatic increase in my heart rate, because I remember a distinct, disconcerting feeling of drifting helplessly and not being in control during those first few seconds of free flight. This sensation continued until using the translational and rotational control sticks, I began firing the MMU thrusters to orient and position myself.

At that point, the responsiveness of the MMU to each command reassured me that this machine was going to take me exactly where I wanted and, incidentally, precisely in the manner for which the simulators back on Earth had prepared me. From then on, the flight was "comfortably exciting."

A real concern during the MMU flight was one that flyers of aircraft and spacecraft have always been worried about—running out of gas!

DO NOT TOUCH
SURFACE

The nitrogen gas that propels the MMU is stored under pressure in two scuba-like tanks. When full, these tanks register about 3000 pounds per square inch of pressure on two small gauges visible to the MMU pilot. By the way, these are the "only" gauges or instruments on the MMU—everything else is left up to the basic seat-of-your-pants flying and plenty of training and simulation on the ground before launch. As you use up nitrogen, the gauge pressure decreases, giving you some feel for your consumption rate as well as fuel remaining.

Because of concerns about MMU thruster gas impinging upon the Weststar VI satellite and disturbing its orientation as I drew near, I was purposely flying the MMU in a control mode that prevented any forward firing thrusters from being used. By keeping the MMU pointed at the satellite during the final approach phase and docking, I could be assured of not disturbing the satellite.

Astronaut Dale Gardner begins his docking approach

Astronaut Dale Gardner closes in on the satellite

With a tool called a "Stinger" aka an Apogee Kick Motor, Astronaut Dale Gardner captures and secures the spinning satellite,

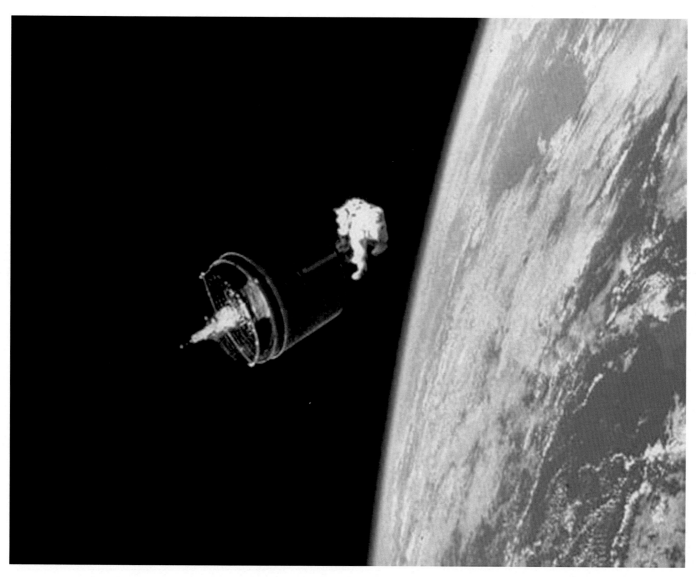

Using the MMU's reverse thrusters Astronaut Dale Gardner stops the spinning satellite's motion, then, slowly, carefully, he delivers the satellite back to the Shuttle's cargo bay.

Astronaut Dale Gardner prepares to disengage himself and the MMU from the satellite as partner and co-astronaut Joseph P. Allen steadies and secures it.

Unfortunately, this mode makes the MMU quite a gas hog. A minute or so from docking I glanced at the gauges and saw that I was already close to my "bingo" of fuel pressure—that reading which by established mission rules would require me to return immediately to the Shuttle! I would have loved to have been able to pull up to a self-serve gas pump and fill'er up with unleaded.

Lacking that possibility, I began to conserve gas with even more vigor than previously. These conservation measures paid off, for at the end of the MMU flight I stowed the MMU in the payload bay with both gas gauges right at the minimum fuel reading (okay, maybe just a bit below minimum—but not much!).

When you ask most astronauts if they experience fear during space flight, many will answer that they do—but it is fear of failure or error, not fear for their lives or well being. In the case of the Shuttle, the astronauts are key players in a multi-hundred million-dollar mission aboard a multi billion-dollar spaceship being watched, directly or indirectly, by most of the inhabitants of planet Earth. Not a good time to mess up!

During my MMU flight, I had an experience that almost put me in that situation. After docking with the Westar VI, using a device fitted to the front of the MMU called the "stinger," I stopped the satellite from spinning by firing my MMU thrusters. Normal procedures then called for me to disconnect by pulling a handle that released only the tip of the stinger, leaving it stuck in the satellite. I would then fly back to the payload bay where Joe Allen would remove the bulky remainder, the stinger from the front of my MMU, and stow it on a mounting rack. This is how we trained countless times on the ground before launch. Because of equipment problems on our first space walk, however, we had changed our plans for the disconnect procedure. I was, under the new plan, supposed to disconnect the MMU entirely from the stinger, leaving the whole device attached to the satellite in its entirety.

This change was necessary because Joe was not available to help me take the stinger from the MMU and stow it; he was on the end of the robot arm holding the satellite in his hands! Well, when Shuttle Pilot Dave Walker radioed to me that it was time to disconnect from the satellite,

I entered my Pavlov dog mode and did exactly what I had trained to do—I pulled the normal disconnect handle. Within a few seconds, I realized that I had made a mistake, followed shortly by that sinking feeling in my stomach that usually accompanies such realizations. As I flew back to the payload bay with the stinger still attached to the MMU,

I remember making a radio call to Dave in Discovery saying, "We screwed up!" I guess the "we" was a weak way of complaining to Dave that he should have reminded me to use the new procedure. But in actuality I knew there was only one person to blame. Not a word was said back to me from Discovery's crew cabin. I could see faces looking out the windows at me and their expressions said it all— "Gardner, you got into this mess. Now, we're going to be quiet and watch while you get out of it."

Sometimes even the weakest of minds works quickly in the extreme, and mine turned out not to be an exception to the rule. I realized that I had to come up with a way to remove the stinger from the MMU and tie it off to Discovery, all while still flying the MMU in a hover in the payload bay. Although this sounds like a four-handed job, weightlessness and the stable flying qualities of the MMU made it possible.

I flew up next to a fixture in the payload bay that was sticking out conveniently, hovered within arm's reach of it, and then put the MMU into an "attitude hold" mode so that it wouldn't turn while I took my hands off the control sticks and went to work. I carefully disconnected the stinger and, using a fastening cord called a tether that was stored on my suit, tied the stinger to the protruding arm.

The MMU

Astronaut Dale A. Gardner

Occasionally, I would have to reach down to the MMU translational controller to move back in position, but it ended up being a relatively simple task. The stinger subsequently waved around on the end of the tether until later, after I had doffed and stowed the MMU, I was able to return to the stinger and properly store it. As you might suspect, I did not chose to make a big deal of this episode during the post-flight debriefings and reports, and NASA was kind enough not to ask too many questions!

As I flew out to the Westar VI satellite and attached myself to it, I had very little opportunity to observe the Earth, Discovery or the vast blackness around me. My concentration was, as you might expect, focused on flying and docking tasks. However, once attached to the satellite and having stopped its rotation, I had an unforgettable opportunity to relax and absorb the universe around me. This came about because it took some time for Rick Hauck to fly Discovery over to my location and Anna Fisher to prepare to take the satellite from me using the Shuttle's robot arm. During that delay, I had nothing to do but wait, so I turned off the automatic attitude hold feature of the MMU to save fuel and relaxed. Not attempting to hold any particular orientation with respect to the Discovery, I eventually rotated to a position from which I could see neither the Shuttle nor the Earth, only space.

Through the radio in my spacesuit, I could hear my fellow crewmembers in Discovery, my space walking partner Joe Allen still in the payload bay and the folks at Houston talking back and forth. I put those discussions into the background because they were not meant for me, and as I did so I had a feeling that it would take only an exceedingly small leap of imagination for me to believe that I was now alone in the cosmos.

This feeling was not disconcerting or frightening but rather calming. And I can remember at the time attempting to analyze the reason. At first, I thought it was the result of having just completed a difficult task, combined with the knowledge that the Shuttle was "really" out there someplace and would come to get me. But somehow I knew that was not the answer; nor was it a spiritual or religious response. The explanation that seemed to fit best was that I was not in a strange or forbidding place at all, but in a place where I --" I" in a sense of being a member of the human race—was meant to be. I thought of those who claim that man should not fly because he was not given wings and have conjured up, I am sure, similar analogies for space travel. As I looked at my space suit and the MMU, however, I knew that we "are" meant to travel away from the Earth because we have been given the curiosity, the intelligence and the will to devise the means and the wonderful machines – such as those that now enclosed me-- to permit such adventures.

I was pulled from my thoughts and the vista surrounding me when from Discovery I finally heard Dave Walker's voice over the radio say, "Dale, we're in position to take the satellite. Give me a slow pitch up now on the MMU and stop it on my call." It was time to go back to work.

Astronaut Joseph P. Allen (seen in visor) snaps
Astronaut Dale Gardner as he raises up a humorous sign.

"Anyone interested in a ten-ton satellite?"

On the other side of the globe, the Soviets who
were the first to place a man in space

(Yuri Gagarin),

the first to place a woman in space

(Valentina Tereshkova),

and the first to accomplish a tethered spacefloat

(Alexi Leonov),

were about to debut their own version of the
Buck Rogers jetpack.

The Soviet MMU
(SPK)

Designs for a Soviet Manned Maneuvering Unit or SPK, which translates into "instrument for the movement of cosmonauts," originated during the early 1960's after Yuri Gagarin had orbited the Earth and an orbiting Soviet Space Station was seen as the next logical step. The Remova, the first design for an SPK, was a tin can type of design that the cosmonaut climbed into. Its primary function as with the American MMU was for Soviet cosmonauts to easily work outside of their spacecraft in zero gravity. Unlike the untethered MMU designs, however, the Remova was designed to be tethered.

Continued research and development was needed, but, with the first Soviet space station Salyut 1 still a decade away, the resources for their SPK were redeployed.

The present day SPK has many nick-names. The most common is "Space Bike." It was developed in 1985 at the Zvezda (Star) factory in Moscow under Gai Illch Severin, the factory head who also worked on the Remova in the 60's. The SPK is larger and bulkier than its American counterpart is. It extends from the helmet to below the knees. The intense cold of outer space prompted the engineers to shield the SPK with a layer of thermal blankets.

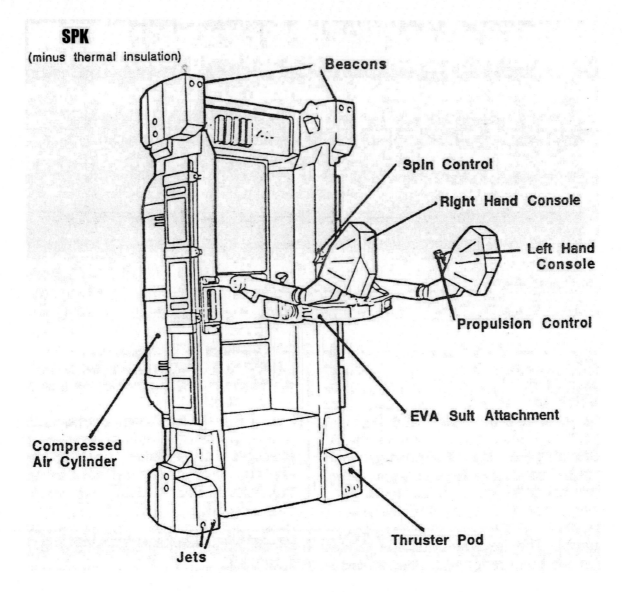

SPK
(minus thermal insulation)

Beacons

Spin Control

Right Hand Console

Left Hand Console

Propulsion Control

EVA Suit Attachment

Compressed Air Cylinder

Thruster Pod

Jets

The Soviet Manned Maneuvering Unit (SPK)

To board the SPK, which is attached to the spacesuit, the Cosmonaut swings open the back door panel, enters, secures the door panel, raises the control arms into the flight position and powers up. As with the Remova, the SPK is designed for orbital service and repair and was assigned to their second Space Station Mir (Peace). Following the Remova's design, the SPK is tethered to Mir for safety. To spacewalk, the Cosmonaut dons on his SPK, depressurizes Mir's Kvant 2 airlock, opens the hatch and spacewalks outside.

Cosmonauts Flying Like Buck Rogers

Aleksandr Stepanovich Viktorenko

Aleksandr Aleksandrovich Serebrov

The SPK weighs 440 pounds, has a top speed of 30 meters per second and is powered by compressed air. The SPK has 32 thrusters (16 primary and 16 backups) mounted in four t-shaped thruster pods. The SPK has two separate sets of controls, one for moving forward and backwards, and another for moving up and down, left and right. The spin control is mounted on the right-hand console, and the propulsion control on the left-hand console. When tethered, the SPK can move 60 meters away from the space station; untethered, it can reach 100 meters distance.

Flying Alongside Mir

Six years after Bruce McCandless became the first Earthman to go Flying like Buck Rogers the Soviets followed suit. In September 1989, the Soviet Soyuz TM-8 spacecraft docked with the Mir Space station and brought the SPK and Cosmonauts Alexander "Sasha" Serebrov and Alexander Viktorenko to fly it.

On February 1st, 1990, piloting the SPK, Sasha stepped out into space and took his first spacewalk. Tethered to the Mir, Sasha began his flying like Buck Rogers miles above the Soviet and Chinese border.

February 5th, Alexander Viktorenko took his turn with the SPK and became the second cosmonaut to spacewalk. Inspecting the Mir space station's exterior for the effects of constant cosmic and solar radiation was each Cosmonauts primary task but this did not bar Alexander, whose image was being broadcast live throughout all of the USSR, from having some live, on air space flight fun.

Alexander "Sasha" Serebrov becomes the first
Soviet Cosmonaut to Fly Like Buck Rogers

To the surprise and joy of the young Soviet's watching his spaceflight from far below, and to conclude the first SPK Buck Rogers style flight, Alexander performed a Cosmonaut first:

a full 360-degree outer space cartwheel!

A New SPK

During the early 1990's the SPK's engineer Gai Severin envisioned a new SPK. His vision then was to add increased impulse power, an improved actuator, and a new telemetry system. With a telemetry system added to the SPK, Severin reasoned, a space dispatcher on the Mir could coordinate the activities of two or three cosmonauts working outside the Mir simultaneously and return an incapacitated cosmonaut back to the Mir through an on-board remote control system.

Cosmonauts Viktorenko and Serebrov smile after completing their history making spacewalks

Cosmonaut Alexander Viktorenko begins the
first-ever space cartwheel.

The S.A.F.E.R.
A Jetvest styled mini MMU

Designs for a new MMU were also on NASA's agenda. The S.A.F.E.R. (Simplified Aid for EVA Rescue) development began in 1992. In designing the S.A.F.E.R., a smaller non-life supporting version of the original MMU, aeronautical designers sent an honorable salute to Inventor Thomas Moore. The SAFER, with a chest-mounted control unit (vs. control arms), is a modern version of Thomas' Jetvest.

Designed as an Astronaut's life jacket, the SAFER's primary use is to provide rescue mobility for an astronaut in danger.

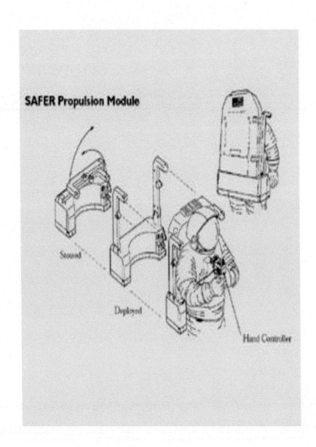

Astronaut Mark Lee was the first to try the rescue
capabilities of the SAFER mini MMU

Astronaut Mark C. Lee appears to be helplessly falling away.
Then using his SAFER mini MMU he safely rockets back
to the Shuttle

With the advent of the SAFER mini MMU and a proposed new SPK, it was decided to retire the original MMU's. The MMU flown by Bruce McCandless II can be seen today at the Smithsonian's Steven F. Udvar-Hazy Center. The MMU flown by Dale Gardner is on display at the Huntsville Space and Rocket Center. A copy of the SPK used by Alexander Serebrov and Alexander Viktorenko is periodically shown at science museums across the country (the original was destroyed upon re-entry).

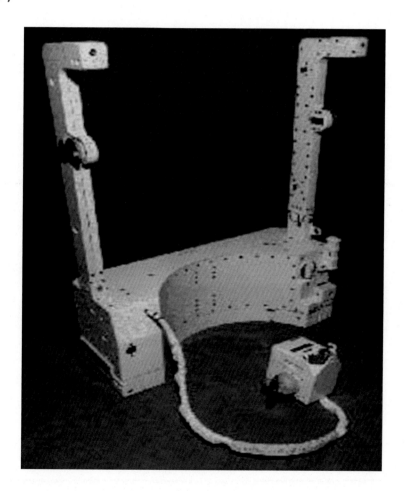

The Soviet MMU (SPK) on display at the Boston Museum

The American MMU on display at The Kennedy Space Center

End Page

Having read this book you've come to know the very special people who inspired by the visions of Buck Rogers turned the dream of solo flight into reality.

If you've a dream to fly just like Buck Rogers, pursue your dreams. Seek the knowledge, meet the challenge and never cease to persevere!

Harold "Pete" Graham – first to fly like Buck Rogers

Putting Your Dreams to Flight:

Astronaut Training
by
Irene Willhite,
USSRC Curator and Archivist

The U.S. Space & Rocket Center, in Huntsville, Alabama, is recognized as one of the most comprehensive U.S. manned space flight hardware museums in the world. Our facilities include Spacedome Theater, Rocket Park, the Education Training Center, which houses NASA's Educator Resource Center, and more.

The Davidson Center for Space Exploration is like no other in the country. In its 476 foot long, 90 foot wide and 63 foot high stucture, suspended 10 feet above the floor, is a national historic treasure, the mighty Saturn V, restored to Apollo era readiness.

Attached to the U.S. Space & Rocket Center are the US Space Camp and Aviation Challenge, unique and remarkably successful links between outer space, life on Earth, and the training of aircraft.

The age groups in Space Camp and Aviation Challenge are nine years old to eighteen. There is also a Parent-child camp, A Corporate Camp that promotes team building in the work place, and, in the fall, Adult Space camp. Campers learn about rocket engines, gyroscopes, guidance and control systems, communication links, aerodynamics, trajectories, and life support systems among many other elements of team-work and the space program. Simulated flights in the cockpit of the shuttle or aircraft where the crew is presented anomalies are exciting events the campers deal with.

The campers, young and old, fulfill their dreams of 'flying in space' as they are attached to the Robotic Arm and float above a satellite while they repair it. Buck Rogers would be envious, or, better yet proud of their accomplishments. Not only do the campers experience actual classroom, hands-on missions, they experience the feelings of being free of the bounds that tie us to Earth. The 'classrooms' are actual training modules once used by NASA in the early days of ISS astronaut training. Every effort is made by the camp staff to give the campers the actual flight experience. 'Free flight' on the Robotic Arm, weightlessness in the 'tank' where they construct ISS elements under water, experiments in the training modules, indeed, allow the campers to feel like a modern day Buck Rogers!

Putting Your Dreams to Flight:

Aviation
by
American Airlines Captain Jeff Dill

For some of us, flying like Buck Rogers means piloting a plane. In fact, it was his comic strip that introduced jet planes, a mode of aviation unheard of in 1929. If airline aviation is your dream then read along with me.

The career path to airline Captain can take many forms, and many times a mentor helps the student along the path. A mentor can be a member of the family, a neighbor, a teacher or even a professional you decide to write to, like me. Today, you can even find a mentor on an aviation studies DVD.

My mentors were my dad, an aircraft mechanic and my uncle Pete a pilot. My uncle Pete you have come to know as Harold Graham the Rocket Ranger and first to fly a rocketbelt. From my dad I learned the nuts and bolts of an aircraft and largely to "Uncle" Pete's contagious influence, I was never far away from one.

So what does it take to be a pilot? For piloting at any level, an absolute requirement is *determination*. Flying can be broken down into a series of basic maneuvers and concepts that build upon themselves and can be mastered with enough *determination*. And surely there will be challenges and obstacles; some will seem like brick walls. If you have *determination*, nothing will stop you. What about the cost?

Pilot lessons are expensive, there is no denying that. When I was young I learned that the best way to solve the money problem was to become a military pilot; so I set my sights on college ROTC. Civilian or military, most pilots hired by the major airlines have at least a four year degree, and it can be in any major. If smaller craft and private piloting are your preferences, a part-time job can help pay for those lessons. And there are associations that provide scholarships as well. The Aircraft Owners and Pilots Asso. (AOPA) is one of the best. They are truly there for the student pilot and can point you to everything from flight training to Scholarships, Are you ready to begin? You can start by following my Uncle Pete's advice:

If you've a dream to fly just like Buck Rogers, pursue your dreams. Seek the knowledge, meet the challenge and never cease to persevere!

I certify that on August 19, 1962 at St. Mary's Municipal Airport [OYM], PA - I gave AA Captain Jeffrey Dill his first plane ride.

Harold Graham

January 28, 2002

Acronyms

AMU = Astronaut Maneuvering Unit

EVA = Extra Vehicular Activity

SRLD = Small Rocket Lift Device

ZGMB = Zero Gravity Maneuvering Belt

LR = Lunar Rocketbelt

SMU = Single Maneuvering Unit

DMU = Dual Maneuvering Unit

HHMU = Hand-held Maneuvering Unit, aka Zip Gun

JSC = Johnson Space Center

NASA = National Aeronautics and Space Administration

MCC = Mission Control Center

ASMU = Astronaut Stabilized Maneuvering Unit

MMU = Manned Maneuvering Unit

T-PAD = Trunion Pin Attachment Device

SPK = Sredstvo Peredivizheniya Kosmonavtika

S.A.F.E.R. = Simplified Aid for EVA

Aerospace Study Resources 1
compiled by
Kathleen Lennon Clough
Aerospace Historian

Classic Aerospace Readings:

Your Future in Space: the U.S. Space Camp Training Program (1986) by Flip McPhee, Penelope McPhee, Raymond Schulke, Debra Schulke.

Rocketbelt Pilot's Manual: A Guide by the Bell Test Pilot (Apogee Books Space Series) (Nov. 1, 2009) by William P. Suitor

Space Exploration (DK Eyewitness Books) [Hardcover] by Carole Stott

Cool Careers for Girls in Air and Space (2001) by Ceel Pasternak

Astronauts: The First 25 Years of Manned Space Flight - Hardcover (Aug. 1988) by Bill Yenne

First Space Encyclopedia (DK First Reference) by Caroline Bingham (Jan 21, 2008)

NASA: The Complete Illustrated History by Michael Gorn and Buzz Aldrin (Mar 2008)

Russia's Cosmonauts: Inside the Yuri Gagarin Training Center (1986) by Rex D. Hall

One Giant Leap: Apollo 11 Remembered by Piers Bizony (May 1, 2009)

Manned Spaceflight (An Explorer's Guide to the Universe) by Erik Gregersen (Dec 20, 2009)

After Sputnik: 50 Years of the Space Age by Martin Collins (Mar 27, 2007)

Aerospace Historian Kathleen Lennon and her two young Rocket Ranger grandsons

Aerospace Study Resources 2
compiled by
Peter Gijsberts
Aerospace Historian

Classic Solo-flight Articles:

The Jetvest

"Jetvest Invented Here Years Ago" *The Redstone Rocket* (January 25, 1984).

"Redstone Retiree Sees Jetvest Invention Commemorated on U.S. Postage Stamp," *The Redstone Rocket* (February 3rd, 1993).

The Jumpbelt

Eliot Tozer, "Man's First Leap Toward Free Flight," *Popular Science* (December, 1958).

"New Aircraft Devices: Flying Belts," *Industrial Research* (Feb. - Mar,. 1960).

The Aeropak

"The Aeropak," *The Aerojet Booster*, vol.III (May, 1959);

"Aerojet &Thiokol Develop Rocketpacks," *Aviation Week* (October, 1959).

The Rocketbelt

Robert Courter and James Joseph, "I Fly The Man Rockets," *Popular Mechanics* (October, 1964)

Tom Huntington, "Leaping Rockets," *Air & Space Smithsonian* (June/July 1987)

Nelson Louis Olivo, Hal Graham, "The Real Rocketeer"
Final Frontier (Jan. – Feb.,1992)

E.C. Krupp, "Backpack to the Future." *Sky and Telescope*
(April, 2008)

The Bell Flying Belt (Jetbelt)

"Bell to Test New Flying Belt for the Military,"
Aviation Week & Space Technology (July 22, 1968).

Robert Courter, "What It's Like to Fly the New Jet Belt,"
Popular Science (Nov., 1969).

Man Without Wings, *Men of Action* vol.10, no. 4 (1970);

Jet Shoes and Astronaut Maneuvering Units

Jet Shoes Studied as Astronaut EVA Aid,"
Space Technology International (October, 1967).

Covault, Craig. "Skylab Aids Design of Maneuvering Unit."
Aviation Week and Space Technology 100 no. 22 (1974):42-47.

Gregory P. Kennedy, "Jet Shoes and Rocket Packs:
The Development of Astronaut Maneuvering Units,"
Space World (October, 1984).

MMU

Overbye Dennis. "In the Armchair of the Gods."
Discover (April, 1984):14-18.

Keith Wilson, "Flying Free," *Spaceflight* vol.28
(Feb., May, 1986).

Alcestis R. Oberg, "Solo: Joy Riding, 150 Miles overhead,"
Final Frontier vol. no. 2 (June, 1988).

Dale Gardner , "Stepping Out for a Walk."
Final Frontier Magazine, 1992

Soviet SPK

Craig Covault, "Cosmonauts Fly Maneuvering Unit While Tethered to Mir Space Station," *Aviation Week and Space Technology* (Feb.12, 1990).

S.A.F.E.R.

James McKenna. "Rescue Device Shines in Untethered Tests." *Aviation Week and Space Technology* 141, no. 13 (1994):25-26

Peter Gijsberts
Aerospace Historian

Aerospace Study Resources 3
compiled by
Ethan Ornstein
Aerospace Historian

Smithsonian National Air and Space Museum
http://www.nasm.si.edu/

U.S. Space Camp
http://www.spacecamp.com/

Kennedy Space Center
http://www.nasa.gov/centers/kennedy/home/index.html

Houston Space Center
http://spacecenter.org/

Mission Control Center
http://spaceflight.nasa.gov/shuttle/reference/mcc/mcc.html

Challenger Center for Space Science Education
http://www.challenger.org/

Astronaut Central
http://www.astronautcentral.com/

Space Explorers
http://www.space-explorers.com/

American Institute for Aeronautics and Astronautics
http://www.aiaa.org/

National Space Society
http://www.nss.org/

British Interplanetary Society
http://www.bis-space.com/

Classic Aerospace U-Tube Titles

1. First Man to Walk in Space Alexi Leonov (March 18, 1965)
2. First Spacewalk; Gemini 4 - Narrated By Ed White (June 3, 1965)
3. Bell Aerospace Rocket Belt
4. Manned Maneuvering Unit (MMU)
5. Bruce McCandless in Orbit
6. Космические байкеры (Space Bikes aka SPK).

Classic Aerospace DVD's

1. The Right Stuff
2. The Dream is Alive
3. For All Mankind
4. When We Left the Earth
5. From the Earth to the Moon
6. Apollo 13
7. In the Shadow of the Moon
8. Failure is Not an Option
9. Magnificent Desolation
10. Spacecamp

Aerospace Historian Ethan Ornstein, Harold Graham &
Nelson Louis Olivo

Nelson's Inspiration

Inspiration for these 2 volumes came from many sources. The first was in 1965 when as a child I visited the NYC World's Fair and saw Robert Courter fly around the Unisphere Globe - just like Buck Rogers;

Seeing Harold "Pete" Graham on *What's My Line* as he told Johnny Carson and the panel that he flew just like Buck Rogers;

Watching Robert Courter teach Johnny Carson how to fly Just like Buck Rogers;

At Tyler Camera in 1980 where I met a genius inventor named Nelson Tyler who built his own rocketbelt to fly – just like Buck Rogers;

my reviving the Tyler rocketbelt demonstrations with Promoter Clyde Baldschun and Rocketeer Peter Kedzierski so once again a man could fly – just like Buck Rogers;

watching Astronaut Dale Gardner on TV rescuing a satellite and flying – just like Buck Rogers - a recollection he wrote for this book and was published by *Final Frontier;*

Getting the thumbs up from Astronaut James "Ox" van Hoften to include his recollections on the Solar Max rescue and of his own flying like Buck Rogers.

At a Columbia University writing class where I was asked to write about a life-long passion which of course was flying – just like Buck Rogers;

Visiting with Olympic Rocketbelt Pilot William "Bill" Suitor at his upstate home in 1989 and reliving his days of flying just like Buck Rogers;

Befriending Rocketeers Gordon Yaeger and Robert Courter whose recollections are a special highpoint of this book. And of course my 20+years of friendship with his "eminence" Harold "Pete" Graham whose story we wrote together culminating in the "Real Rocketeer" article, also published by *Final Frontier.*

INDEX

A

Aerojet General Corporation, 22-24
Aerojet Pogo, 22
Aeropak, 22-24
Allen, Joseph, 116-118
AMU (astronaut maneuvering units), 76, 79-80, 94
ASMU (astronaut stabilized maneuvering unit), 93, 153

B

Baldschun, Clyde, 41
Bell Aerosystems, 25, 27-28, 30-31, 51, 57
Bohr, Alexander, 18-21
Bond, James, 48
Burdett, Harry, 18-21

C

Cernan, Eugene, 79-80
Collins, Michael, 79
Courter, Robert, 3, 5, 41, 48, 156-57
Crevar, Mitch, 31

D

Deboy, Marve, 31
Deering, Wilma, 12
Dill, Jeff, 151-152

E

EVA (extravehicular activity), 75, 79-80, 99, 110

F

Fisher, Anna, 116, 131
Flying belt (Thiokol), 18-21

G

Gagarin, Yuri, 73, 134
Ganczak, Eddie, 31
Gardner, Dale, 116-17, 119-133, 145

George, Millie, 31, 35, 57
Gijsberts, Peter
Glines, Alan, 81-94
Gordon, Dick, 80
Graham, Harold "Pete," 27-57, 148,160

H

Hart, Terry, 113
Hauck, Rick, 131
Hess, Cliff, 95
HHMU (handheld maneuvering unit), 77, 79-80, 93-94, 97
Huntsville Space and Rocket Center, 145, 149-150
hydrogen peroxide, 22, 34, 80

I

Imagineering, 11

J

Jet flying belt (Bell). 63-69
Jet shoes, 93-94

K

Kedzierski, Peter, 41-43, 65
Kedzierski, Stan, 20
Kennedy, John F., 9, 11, 14, 38-39
Kolonel Keds, 44
Kreutinger, Ernie, 30-31

L

Lee, Mark, 143-44
Lennon, Tom, 3, 5, 28, 31
Lennon-Clough, Kathleen, 155
Leonov, Alexi, 74, 134
Lovell, Jim, 89
LTV (Ling-Temco-Vought), 76

M

McCandless, Bruce "Buck," II, 95-107
Mission Control Center, 81-88
MMU (American manned maneuvering unit), 29, 95-108, 110, 112-13, 116-17, 119-125-132

Moore, Thomas, 3, 10-16, 25
Moore, Wendell F., 5, 25, 27-28, 30, 35, 57, 75-76

N

Nelson, George "Pinky," 110-112
Nowlan, Philip, 7

O

Olivo, Nelson, 51, 57, 160, 168
Ornstein, Ethan 160

P

Pabst, Glen, 31
Palapa, 115
Parkin, Charles, 18-19
Peoples, Richard, 23-24
Powel, Jim, 25
propellants, 22, 67, 85, 94, 96, 99

R

Reagan, Ronald, 105
RMS (remote manipulator system), 113
rocket belt, 27, 29, 30, 32, 34, 35, 40-47
rocket pack, 30
Rogers, Buck, 7-8, 12, 15, 21, 26-27, 30-31, 35, 41, 49, 63, 67, 70, 72, 76-77, 82, 87, 95,107, 134, 139, 150-52

S

SAFER (simplified aid for EVA rescue), 143-45
Serebrov, Alexander "Sasha," 137-40, 145
Severin, Gai, 135, 140
Shepard, Alan, 73, 87
Sherry, Norm, 31
Skylab, 80, 88, 92-93
SMU (single maneuvering unit), 76
Solar Max, 110, 112-13
SPK (Soviet Manned maneuvering Unit), 135-141, 145-46
SRLD (small rocket lift device), 19, 21-22, 24-25
Stewart, Robert "Flash," 106-108
Suitor, William "Bill", 5, 41, 48, 52-53-57

T

Thiokol Chemical Corporation, 18-19, 21
Thompson, Leo, 31
Thunderball, 48
TPAD (trunnion pin attachment device), 112
Trudeau, George, 22-24
Tyler, Nelson, 49-50
Tyler Rocketbelt, 49-55

U

US Space Camp, 15-16, 149-150

V

van Hoften, James "Ox," 110, 113-114
Viktorenko, Alexander, 137-138, 140-41, 145
von Braun, Wernher, 10-12, 14, 25, 27

W

Walker, Dave, 128, 132
Westar, 115
White, Edward, 77, 79
Wiech, Ray, 20
Willhite, Irene, 149-150,
Williams International, 48, 67
Witset, C. E., 95

Y

Yaeger, Gordon, 41, 48, 58-62
Yeager, Chuck, 27

Z

ZGMB (zero gravity maneuvering belt), 75, 153
zip gun. See HHMU (handheld maneuvering unit)

Photo Credits

Aerojet General	23-4
Alan Glines	90-91
Bell Aerosystems	25, 41-44
Bill Clark	3
Bill Suitor	56, 63-68
Boston Museum	146
Cathleen Lennon – Clough	28, 155
Dale Gardner	119, 121-26, 133
Ethan Orenstein	45, 47
Gordon Yeager	58-9, 61-2
Harold Graham	29-34, 36-9, 148
Irene Graham	71, 160
Huntsville Space & Rocket Center	77
John Guerriero	57
Jeff Dill	152
Kennedy Space Center	147
Kim Garrett	51
Johnson Space Center	106, 108, 111, 113-14, 118, 130
LA 84 Foundation	53-55
Mark Wells	14, 167
MGM (United Artists)	48
NASA	73-4, 78, 80-1, 83, 86-7, 92-4, 96, 98, 100-4, 107-10, 112, 115-17, 129, 142-45
Nelson Louis Olivo	51
Nelson Tyler	49-50
National Newspaper Syndicate	7
Peter Gijsberts	158
RKK Energia	137,139-141
Robert Courter	69, 76
Ronald Reagan Museum	105
Smithsonian Institute Library/Archives	4, 23-4,
Thiokol	19-21
Thomas M. Moore	6, 9-13, 15-17
US Army Transportation Museum	40
US Space Camp	149-50

Thank you:

Gloria Rangelli, Frank Lantigua, Evelyn Rosario, Clyde Baldshun, Nelson Tyler, Peter Kedzierski, Lucy Negron, Joy Chapman, John Guerriero, Richard Dutchik, Dr. Gail Leggio, Thomas M Moore, Bell Aerospace, US Army Transportation Museum, S.U.N.Y. Buffalo - Engineering , Queens Museum of Science, Boston Museum of Science, Harold "Pete" Graham, Irene Graham, Millie Weaver, Millie George, Edward Ganczak, Robert Courter, Carolyn Moore-Baumet, William Bill Suitor, Barbara Kushner, Kathleen Lennon-Clough, Williams International, Rex Hall, James Oberg, Archives @ U.S. Space Camp, Kennedy Space Center, Gordon Yaeger, Thiokol, Frank Winter @ Air and Space Smithsonian Library, Aerojet General, John Koch, Brett Hoffstadt, Peter Gijsberts, James "Ox" van Hoften, Ethan Orenstein, Dale Gardner, Trenton Free Public library, Lori Novis, Christos Nicola, Kathi Eckert, Martin Marietta, William Clark, Dr. Larry Kroah, Irene Percelli, Alan Glines, Mark Wells, Judy Wells, Irene Willhite, James Willhite, Catherine Smith, Ming Zhu Chen, Alfonso Alejo, Dr. Gerard Dapena, Crystal Schroder and Mike Gentry @ Johnson Space Center, Michael Salmon @ LA84 Foundation, Keith Carpenter & Calvin Blackshear, Kate Igoe, Jim Stanmore, Mirian Montalbo, Jane Guatno, Martin Kintanar.

The official Rocket Ranger patch

A "Thank You "Thumbs Up" from Rocket Ranger
Harold "Pete" Graham for buying our books.

L-R: Nelson Louis Olivo, Judy Wells, Millie Weaver,
Harold "Pete" Graham, Mark Wells

Printed in the United States
By Bookmasters